Tropical Birds
of Australia

Tropical Birds
of Australia

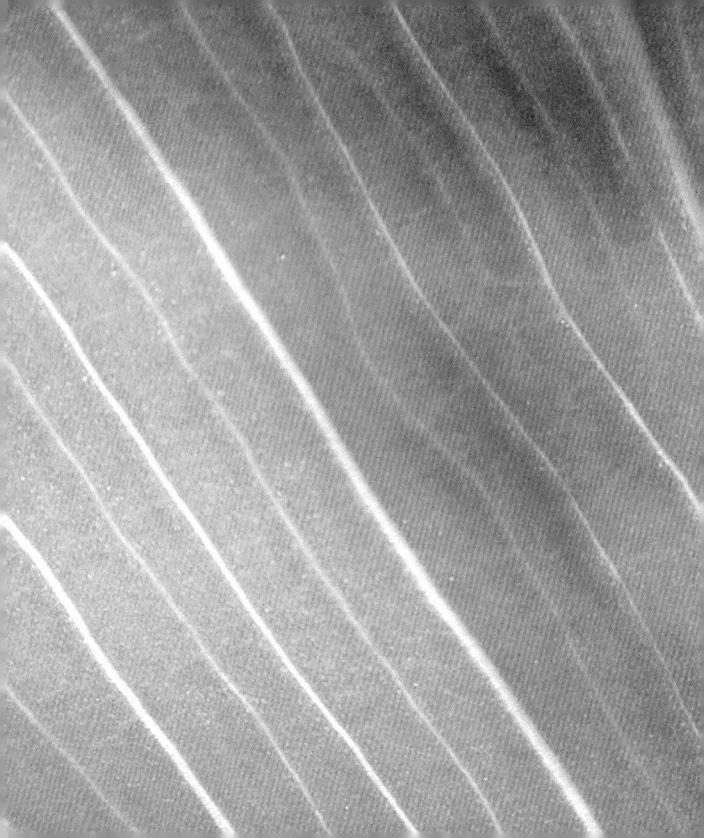

INTRODUCTION

The tropical birds of Australia are some of the most beautiful, colourful and charismatic in the world. They inhabit a range of environments, from dense tropical rainforests to mangroves, wetlands, grasslands and parks. All have one thing in common: they can be found in 'tropical' Australia – surrounding the Tropic of Capricorn that runs across the centre of the country – but may also be found further afield.

The selection of tropical birds in this book represents a cross-section of the thousands of amazing species that can be found in Australia. Some are familiar icons that can be readily spotted in many parts of the country, such as the rosella, cockatoo and kookaburra. Others, such as the wompoo fruit dove, magnificent riflebird or Gouldian finch, may be less familiar.

Unfortunately, many of these beautiful creatures are scarce, and as a consequence are listed as vulnerable or endangered by the relevant state authority. The very characteristics that set tropical birds apart – their beauty and rarity – make them prime targets for poaching for sale as pets.

Pressures as a result of habitat loss, climate change and pollution are especially serious as many of the bird species that are threatened can be found nowhere else in the world. The importance of protecting these threatened birds must be recognised for their future to be secure.

The information on each of the featured birds paints a broad picture of where and how these birds live. The geographic range and habitat describe where in Australia the birds can be found, and the types of environments they tend to inhabit. Information regarding size and diet is included for each species. Details of breeding season, nesting habits, number and type of eggs laid, incubation and fledge time describe the diverse habits of the birds as they mate, make a home, and tend to their young.

Although often difficult to describe, the characteristic voice of each bird is also included. And finally there are some general notes that identify any particular characteristics that have not been covered by the preceding headers.

AUSTRALASIAN FIGBIRD
Sphecotheres vielloti

GEOGRAPHIC RANGE: north coast of WA and NT; north and east coast of Qld; east coast of NSW.

HABITAT: tropical rainforest edges and parks.

SIZE: 27–30 cm.

DIET: fruit and insects.

BREEDING SEASON: August to April.

NEST: deep and cup-shaped, suspended by the rim.

EGGS: 2–3 dull greenish eggs with red-brown marks and spots.

INCUBATION: 17–18 days.

FLEDGE TIME: 2 weeks.

VOICE: short sharp yelps, or soft musical calls.

GENERAL: True to its name, the figbird is often found in fig trees. Whereas the female is quite dull in colour, the male is a spectacular green, with red spectacles.

AUSTRALIAN KING-PARROT
Alisterus scapularis

GEOGRAPHIC RANGE: east coast of Qld, NSW and Vic.

HABITAT: forest, farmland, orchards, parks and gardens.

SIZE: 40–45 cm.

DIET: blossoms, fruit, leaf buds, nectar and shoots.

BREEDING SEASON: September to January.

NEST: tree and limb hollows, lined with chewed wood.

EGGS: 3–8 glossy, smooth and white.

INCUBATION: 20 days, mostly by the female.

FLEDGE TIME: around 5 weeks.

VOICE: 'carrak-carrak'.

GENERAL: Like most parrots, the king-parrot mates for life, with the male distinguished by its red head and breast, and the female by its green head, breast and wings.

AUSTRALIAN RINGNECK
Barnardius zonarius

GEOGRAPHIC RANGE: western half of WA; south NT; most of SA; south Qld; central NSW and Vic.

HABITAT: scrubland, woodland and wet forest.

SIZE: male 32–44 cm; female 28–41 cm.

DIET: fruit, nuts and seeds.

BREEDING SEASON: variable.

NEST: tree and limb hollows, lined with chewed wood.

EGGS: 3–8 glossy, smooth and white.

INCUBATION: 20 days, mostly by the female.

FLEDGE TIME: around 5 weeks.

VOICE: mellow whistles.

GENERAL: There are several types of Australian ringneck but they are all green with a distinctive yellow collar and a long tail.

BLUE-FACED HONEYEATER
Entomyzon macleayana

GEOGRAPHIC RANGE: north coast of WA and NT; north and east coast of Qld; most of NSW; north half of Vic.

HABITAT: forests and gardens.

SIZE: 25–30 cm.

DIET: fruit, insects and nectar.

BREEDING SEASON: June to March.

NEST: small cup of twigs, bark and moss.

EGGS: 1–2 pale eggs with darker areas.

INCUBATION: 12–17 days by the female.

FLEDGE TIME: 10–15 days.

VOICE: 'ki-owt'.

GENERAL: The blue-faced honeyeater is a large, noisy and aggressive honeyeater with an olive-green body, white breast, black head and blue facial skin patch.

BLUE-WINGED KOOKABURRA
Dacelo leachii

GEOGRAPHIC RANGE: northern coasts of WA and NT; north and east coast of Qld.

HABITAT: woodland, forests and swamps.

SIZE: 40 cm.

DIET: crayfish, earthworms, insects and reptiles.

BREEDING SEASON: August to February.

NEST: tree hollows.

EGGS: 2–6 white.

INCUBATION: 28 days.

FLEDGE TIME: 3–5 weeks.

VOICE: cackling and twittering.

GENERAL: The kookaburra is one of Australia's most familiar and best-loved birds. The blue-winged kookaburra, with its bright wings, is especially beautiful.

BROLGA
Grus rubicunda

GEOGRAPHIC RANGE: north coast of WA and NT; most of Qld, NSW and Vic.

HABITAT: wetlands, saltmarshes and open grassland.

SIZE: male 105–134 cm; female 77–113 cm.

DIET: crustaceans, frogs, grain, insects, molluscs, seeds and tubers.

BREEDING SEASON: September to May.

NEST: mound of grass, sticks and leaves raised above water level.

EGGS: 2–3 white marked with brown.

INCUBATION: 28–31 days, mostly by the female.

FLEDGE TIME: 8–11 weeks.

VOICE: whooping and trumpeting.

GENERAL: Australia's only crane, the stately brolga has a featherless red head and grey crown. They perform spectacular mating rituals of bows, leaps and high steps.

BUDGERIGAR
Melopsittacus undulates

GEOGRAPHIC RANGE: most of Australia, except the east and north coast.

HABITAT: woodland and grassland.

SIZE: 18 cm.

DIET: seeds.

BREEDING SEASON: June to January.

NEST: tree and limb hollows, lined with chewed wood.

EGGS: 3–8 glossy, smooth and white.

INCUBATION: 18 days, mostly by the female.

FLEDGE TIME: 5 weeks.

VOICE: 'chirrup'.

GENERAL: The budgerigar is a small nomadic parrot beloved by many. They are a distinctive yellow and green with black patterning, and congregate in dense flocks. The colouring of budgerigars varies in domestic varieties.

BUFF-BREASTED PARADISE-KINGFISHER
Tanysiptera sylvia

GEOGRAPHIC RANGE: north-east coast of Qld.

HABITAT: lowland rainforest.

SIZE: male 36 cm; female 30 cm.

DIET: crabs, fish, frogs, insects, lizards, mice, snails, snakes, tadpoles and yabbies.

BREEDING SEASON: October to March.

NEST: tunnels in termite mounds.

EGGS: 2–6 white.

INCUBATION: 21 days.

FLEDGE TIME: 3–5 weeks.

VOICE: rising piping.

GENERAL: The beautiful vivid plumage of the buff-breasted paradise-kingfisher is gold, white, black and purple-blue, with a red beak and a long tail streamer.

CHANNEL-BILLED CUCKOO
Scythrops novaehollandiae

GEOGRAPHIC RANGE: throughout north and east.

HABITAT: tall trees.

SIZE: 57–70 cm.

DIET: baby birds, figs, fruits, insects and seeds.

BREEDING SEASON: July to March.

NEST: uses other birds' nests.

EGGS: resemble their hosts' eggs in size and colour.

INCUBATION: by host species.

FLEDGE TIME: may be forced to leave nest early.

VOICE: hoarse trumpeting.

GENERAL: The channel-billed cuckoo is the largest parasitic cuckoo in the world, with a huge pale bill used to take fruits from the trees, a red iris and eye ring, pale body and grey wings.

CHESTNUT-BREASTED MANNIKIN

Lonchura castaneothorax

GEOGRAPHIC RANGE: north and east coast.

HABITAT: reed banks and grass.

SIZE: 10 cm.

DIET: seeds and termites.

BREEDING SEASON: September to April.

NEST: small, roofed, unwoven, made from grass.

EGGS: 4–6 pure white.

INCUBATION: 12–16 days.

FLEDGE TIME: 3–4 weeks.

VOICE: 'teet'.

GENERAL: The chestnut-breasted mannikin is a small brown finch with a black face and chin, silver beak and yellow tail.

COCKATIEL
Nymphicus hollandicus

GEOGRAPHIC RANGE: most of Australia, except for the south-east coast, Tas., Cape York and central WA.

HABITAT: semi-arid and arid country, mostly near water.

SIZE: 29–32 cm.

DIET: berries, fruits, grasses and seeds.

BREEDING SEASON: variable, mainly April to December.

NEST: hollows in old-growth trees, lined with chewed wood.

EGGS: 1–6 white.

INCUBATION: 20 days, mostly by the female.

FLEDGE TIME: 4 weeks.

VOICE: high, rolling 'cweeree'.

GENERAL: With its slender body and pointed tail the cockatiel closely resembles a parrot. Plumage is mostly grey, with a lemon-coloured forehead, face, throat and cheeks, and an orange ear patch. They are popular pets.

COMB-CRESTED JACANA
Irediparra gallinacean

GEOGRAPHIC RANGE: north and east coast.

HABITAT: swamps, lakes and lagoons.

SIZE: male 20–21 cm; female 24–27 cm.

DIET: aquatic plants, insects and seeds.

BREEDING SEASON: November to May.

NEST: floating herbage in water over 1 m deep.

EGGS: 3–4 marked gold, red and black.

INCUBATION: 21–26 days by the male.

FLEDGE TIME: 6 weeks.

VOICE: 'pee pee pee'.

GENERAL: The comb-crested jacana has a distinctive red, fleshy forehead, and appears to be able to walk on water when seen from a distance as it struts around on long green legs atop floating plants.

CRIMSON ROSELLA
Platycercus elegans

GEOGRAPHIC RANGE: east coast of Qld and NSW; south coast of Vic.

HABITAT: forest, farmland and parks.

SIZE: 35–38 cm.

DIET: seeds.

BREEDING SEASON: August to January.

NEST: tree and limb hollows, lined with chewed wood.

EGGS: 3–8 glossy, smooth, white.

INCUBATION: 19 days, mostly by the female.

FLEDGE TIME: 5 weeks.

VOICE: bell-like whistle.

GENERAL: The colour of the crimson rosella varies considerably, and the crimson plumage tends to be darker on those found in more northerly areas.

DOLLARBIRD
Eurystomus orientalis

GEOGRAPHIC RANGE: throughout north and east.

HABITAT: woodland and watercourses.

SIZE: 26–29 cm.

DIET: insects and lizards.

BREEDING SEASON: September to February.

NEST: unlined hollows of tall trees.

EGGS: 3–4 glossy white.

INCUBATION: 18–20 days.

FLEDGE TIME: 4 weeks.

VOICE: harsh yap.

GENERAL: A migratory bird, the dollarbird visits Australia each year from the north to breed. With its strong body, prominent bill and square tail, it can often be seen perched high and proud on dead trees.

EASTERN ROSELLA
Platycercus eximius

GEOGRAPHIC RANGE: throughout east coast.

HABITAT: woodland, farmland, parks and gardens.

SIZE: 28–33 cm.

DIET: fruit and seeds.

BREEDING SEASON: August to January.

NEST: tree and limb hollows, lined with chewed wood.

EGGS: 3–8 glossy, smooth, white.

INCUBATION: 19 days, mostly by the female.

FLEDGE TIME: 5 weeks.

VOICE: bell-like 'pee-pity'.

GENERAL: Similar to the crimson rosella but even more colourful, the eastern rosella's plumage is a riot of red, yellow, blue and green.

ECLECTUS PARROT
Eclectus roratus

VULNERABLE

GEOGRAPHIC RANGE: tiny area of Cape York, Qld.

HABITAT: rainforest and woodland.

SIZE: male 42–48 cm; female 40–45 cm.

DIET: berries, blossoms, buds, fruit, nectar, nuts and seeds.

BREEDING SEASON: July to February.

NEST: tree and limb hollows, lined with chewed wood.

EGGS: 3–8 glossy, smooth, white.

INCUBATION: 20 days, mostly by the female.

FLEDGE TIME: 4–5 weeks.

VOICE: quavering screech.

GENERAL: The eclectus parrot is unusual in that the male and female birds have completely different colouring: the male a vivid green and the female crimson and blue.

EMERALD DOVE
Chalcophaps indica

GEOGRAPHIC RANGE: north and east coast.

HABITAT: rainforest and mangrove.

SIZE: 23–27 cm.

DIET: fruit and seeds.

BREEDING SEASON: August to June.

NEST: stick nest on ground, rock ledge, tree fork or vines.

EGGS: 1–2 glossy cream/white.

INCUBATION: 14–21 days.

FLEDGE TIME: 3–5 weeks.

VOICE: 'croo, curroo, curroo'.

GENERAL: The beautiful emerald dove has a soft mauve head and breast and iridescent emerald wings. They are found foraging either alone or in pairs.

GALAH
Eolophus roseicapillus

GEOGRAPHIC RANGE: most of Australia.

HABITAT: woodland, open shrubland, grassland and parks.

SIZE: 35 cm.

DIET: crops, grasses and seeds.

BREEDING SEASON: variable.

NEST: hollows of old-growth trees, lined with chewed wood.

EGGS: 1–6 white.

INCUBATION: 28–30 days.

FLEDGE TIME: 7 weeks.

VOICE: 'chri-chri'.

GENERAL: The charismatic and much-loved galah is commonly seen throughout Australia in permanent pairs or large family flocks.

GOLDEN BOWERBIRD
Prionodura newtoniana

GEOGRAPHIC RANGE: tiny area of north Qld.

HABITAT: tropical rainforest.

SIZE: 23–25 cm.

DIET: berries, fruit, insects and seeds.

BREEDING SEASON: September to March.

NEST: cup of twigs, bark and leaves.

EGGS: 1 white.

INCUBATION: 19–21 days by the female.

FLEDGE TIME: 3 weeks.

VOICE: rattles, croaks, mimicry.

GENERAL: The golden bowerbird is the smallest but most visually striking of the bowerbirds, and the male builds a 'maypole' bower up to 3 m tall adorned with flowers, lichen and berries to attract a mate.

GOLDEN-SHOULDERED PARROT

Psephotus chrysopterygius

ENDANGERED

GEOGRAPHIC RANGE: central north Qld.

HABITAT: savannah woodland.

SIZE: 23–28 cm.

DIET: seeds.

BREEDING SEASON: April to August.

NEST: tunnels in termite mounds.

EGGS: 3–8 glossy, smooth, white.

INCUBATION: 20 days, mostly by the female.

FLEDGE TIME: 4–5 weeks.

VOICE: 'preep-preep'.

GENERAL: Sadly an endangered species, golden-shouldered parrots are gorgeous, multi-coloured birds.

GOLDEN WHISTLER
Pachycephala pectoralis

GEOGRAPHIC RANGE: east and south-east coast; south-west WA.

HABITAT: rainforest, woodland and coast.

SIZE: 15–17 cm.

DIET: fruit, grubs and insects.

BREEDING SEASON: August to February.

NEST: neat nest of shredded bark and cobweb, high in the fork of a tree.

EGGS: 2–3 white, with olive or brown markings.

FLEDGE TIME: 2 weeks.

VOICE: rich and melodious.

GENERAL: The female golden whistler is a dull brown and grey colour, but the male is a vivid gold, black and green, with a white chin.

GOULDIAN FINCH
Erythrura gouldiae

ENDANGERED

GEOGRAPHIC RANGE: north and east coast.

HABITAT: open woodland and grassland.

SIZE: 14 cm.

DIET: grasses and insects.

BREEDING SEASON: November to April.

NEST: inside hollow tree branches.

EGGS: 4–6 white.

INCUBATION: 12–16 days.

FLEDGE TIME: 3–4 weeks.

VOICE: 'ssitt'.

GENERAL: Gouldian finches are some of Australia's most spectacularly coloured birds, which has unfortunately led to them being sought after by the illegal bird trade.

GREAT FRIGATEBIRD
Fregata minor

GEOGRAPHIC RANGE: off the coast of the north and north-east.

HABITAT: tropical seas.

SIZE: 86–100 cm.

DIET: crustaceans and fish.

BREEDING SEASON: April to December.

NEST: stick platforms on ground or low vegetation.

EGGS: 1 chalky white.

INCUBATION: 45–50 days.

FLEDGE TIME: 17–21 weeks.

GENERAL: The great frigatebird is an enormous sea bird with a red throat pouch that it inflates during courtship. Frigatebirds are known to harass other birds for food.

LEMON-BELLIED FLYCATCHER

Microeca flavigaster

GEOGRAPHIC RANGE: north coast.

HABITAT: woodland, mangroves and along streams.

SIZE: 12 cm.

DIET: insects.

BREEDING SEASON: August to March.

NEST: small cups of fine vegetation with cobwebs, mosses and lichens in tree forks or horizontal branches.

EGGS: 2–3 green, blue, cream, buff or white.

INCUBATION: 14–20 days.

FLEDGE TIME: 12–22 days.

VOICE: sweet whistling songs.

GENERAL: A tiny robin, the lemon-bellied flycatcher reputedly builds the smallest nest of any bird in Australia.

© Mick Davey / ANTPhoto.com

MAGNIFICENT RIFLEBIRD
Ptiloris magnificus

GEOGRAPHIC RANGE: northern tip of Cape York, Qld.

HABITAT: wet forest.

SIZE: male 34 cm; female 28 cm.

DIET: fruit, insects and seeds.

BREEDING SEASON: September to March.

NEST: large, strong cup.

EGGS: 1–2 white with reddish streaks.

INCUBATION: by the female.

VOICE: 'whit whit'.

GENERAL: The riflebird may have been named after its resemblance to the uniforms of riflemen in the British army, or because its call is similar to the whine of a fired bullet. Of the three riflebird species in Australia – magnificent, paradise and Victoria's – the magnificent riflebird is the largest.

MAJOR MITCHELL'S (PINK) COCKATOO
Cacatua leadbeateri

VULNERABLE

GEOGRAPHIC RANGE: throughout centre and south-east.

HABITAT: mallee, mulga, Murray pine and she-oak.

SIZE: 39 cm.

DIET: berries, insects, nuts and seeds.

BREEDING SEASON: May to November.

NEST: hollow in old-growth tree, lined with chewed wood.

EGGS: 1–6 white.

INCUBATION: 28–30 days.

FLEDGE TIME: 7–8 weeks.

VOICE: two-note quaver.

GENERAL: Perhaps the most beautiful of all cockatoos, the Major Mitchell is a radiant salmon pink, with a scarlet and yellow crest.

NOISY PITTA
Pitta versicolor

GEOGRAPHIC RANGE: east coast.

HABITAT: rainforest and scrubland.

SIZE: 18–26 cm.

DIET: berries, earthworms, fruit, insects and snails.

BREEDING SEASON: September to March.

NEST: bulky, domed, on or close to the ground.

EGGS: 2–4 white, round, marked purple or brown.

INCUBATION: 15–17 days.

FLEDGE TIME: 2–3 weeks.

VOICE: loud whistle – 'walk to work'.

GENERAL: The noisy pitta is called the Devil-devil bird by some indigenous Australians, who believe that it steals their children by luring them into the rainforest with its jewel-like plumage and piercing call.

NORTHERN ROSELLA
Platycercus venustus

GEOGRAPHIC RANGE: north-east WA; north NT.

HABITAT: eucalypt, woodland, scrub, grassland and gardens.

SIZE: 29–32 cm.

DIET: fruit and seeds.

BREEDING SEASON: May to September.

NEST: tree and limb hollows, lined with chewed bark.

EGGS: 3–8 glossy, smooth, white.

INCUBATION: 19 days, mostly by the female.

FLEDGE TIME: 5 weeks.

VOICE: 'pee-pity, pee-pity'.

GENERAL: The northern rosella is a distinctive bird with yellow colouration, blue wings and tail, black patterning and a black cap.

NORTHERN SHRIKE-TIT
Falcunculus whitei

VULNERABLE

GEOGRAPHIC RANGE: north-east WA; north NT.

HABITAT: woodland.

SIZE: 14–16 cm.

DIET: grubs and insects.

BREEDING SEASON: November to March.

NEST: neat, deep nest of shredded bark and cobweb in fork of tree, high above the ground.

EGGS: 2–3 white with olive and brown markings.

INCUBATION: 16–18 days.

FLEDGE TIME: 2 weeks.

VOICE: plaintive whistling.

GENERAL: A small, pretty bird, the northern shrike-tit is the smallest of the crested shrike-tits. Sadly, it is becoming endangered as a result of habitat loss.

PALM COCKATOO
Prosciger aterrimus

GEOGRAPHIC RANGE: Cape York, Qld.

HABITAT: tropical forest and savannah woodland.

SIZE: 56 cm.

DIET: fruit, leaf buds and seeds.

BREEDING SEASON: August to February.

NEST: hollows of old-growth trees, lined with chewed wood.

EGGS: 1–6 white.

INCUBATION: 28–30 days.

FLEDGE TIME: 8–9 weeks.

VOICE: metallic whistle.

GENERAL: The palm cockatoo is a spectacular bird with a prominent crest, black body and red unfeathered patches on its cheeks.

PRINCESS (ALEXANDRA'S) PARROT

Polytelis alexandrae

VULNERABLE

GEOGRAPHIC RANGE: central west.

HABITAT: arid shrubland and trees along water courses.

SIZE: 40–45 cm.

DIET: seeds.

BREEDING SEASON: September to December.

NEST: tree and limb hollows, lined with chewed wood.

EGGS: 3–8 glossy, smooth, white.

INCUBATION: 20 days, mostly by the female.

FLEDGE TIME: 4–5 weeks.

VOICE: prolonged call and chattering.

GENERAL: An incomparably beautiful bird, the princess parrot's colouring is a gentle pastel blue, pink and green.

RAINBOW BEE-EATER
Merops ornatus

GEOGRAPHIC RANGE: most of the country apart from central west.

HABITAT: open country, sand dunes and banks.

SIZE: 22–25 cm.

DIET: insects.

BREEDING SEASON: August to March.

NEST: unlined tunnel in flat or slightly sloping ground.

EGGS: 4–5 dull white.

INCUBATION: 24–25 days.

FLEDGE TIME: 4–5 weeks.

VOICE: melodious trilling.

GENERAL: The rainbow bee-eater's long beak is ideally suited to catching bees, which the bird then beats on a branch before consuming.

RAINBOW LORIKEET
Trichoglossus haemotodus

GEOGRAPHIC RANGE: east and south-east coast.

HABITAT: rainforest, open forest, woodland, heath, gardens and parks.

SIZE: 30 cm.

DIET: berries, blossom, fruit, insects, nectar, pollen and seeds.

BREEDING SEASON: July to January.

NEST: tree and limb hollows, lined with chewed wood.

EGGS: 3–8 glossy, smooth, white.

INCUBATION: 20 days, mostly by the female.

FLEDGE TIME: 4–5 weeks.

VOICE: 'screet, screet'.

GENERAL: Noisy and gregarious, the rainbow lorikeet is a conspicuous, brightly-coloured bird found in large flocks.

RED-COLLARED LORIKEET
Trichoglossus rubritorquis

GEOGRAPHIC RANGE: east and south-east coast.

HABITAT: tropical open forest, gardens and settlements.

SIZE: 30 cm.

DIET: berries, blossom, fruit, nectar, pollen and seeds.

BREEDING SEASON: March to June.

NEST: tree and limb hollows, lined with chewed wood.

EGGS: 3–8 glossy, smooth, white.

INCUBATION: 24 days, mostly by the female.

FLEDGE TIME: 4–5 weeks.

VOICE: screeching, chattering.

GENERAL: The red-collared lorikeet is the northern cousin of the rainbow lorikeet, with similar plumage but a red rather than yellow collar.

RED-TAILED BLACK COCKATOO
Calyptorhynchus banksii

GEOGRAPHIC RANGE: west, north and east.

HABITAT: coastal forest and woodland.

SIZE: 55–60 cm.

DIET: bulbs, fruit, insects and seeds.

BREEDING SEASON: February to December.

NEST: hollow in old-growth tree, lined with chewed wood.

EGGS: 1–6 white.

INCUBATION: 28–30 days.

FLEDGE TIME: 8–9 weeks.

GENERAL: The red-tailed black cockatoo is a magnificent bird widely distributed throughout Australia. However, the population found in south-west Victoria and eastern South Australia is endangered.

RED-WINGED PARROT
Aprosmictus erythropterus

GEOGRAPHIC RANGE: north and east.

HABITAT: subtropical, semi-arid eucalypt, casuarina woodland and mulga.

SIZE: 32 cm.

DIET: blossom, nectar, pollen and seeds.

BREEDING SEASON: variable.

NEST: tree and limb hollows, lined with chewed wood.

EGGS: 3–8 glossy, smooth, white.

INCUBATION: 20 days, mostly by the female.

FLEDGE TIME: 4–5 weeks.

VOICE: brassy 'crillik-crillik'.

GENERAL: A large bird, wary and shy, the red-winged parrot confines itself to the forest canopy, descending only to drink.

RUFOUS FANTAIL
Rhipidura rufifrons

GEOGRAPHIC RANGE: east coast.

HABITAT: wet forest.

SIZE: 15–16 cm.

DIET: insects.

BREEDING SEASON: September to April.

NEST: cup made from fine material, bark and cobweb.

EGGS: 2–3 white to cream with speckles.

INCUBATION: 14–15 days.

FLEDGE TIME: 2 weeks.

VOICE: thin squeaks ('pseet'), twittering song.

GENERAL: With its striking tail, the rufous fantail can perform acrobatic manoeuvres in pursuit of its prey. It is an active bird, often seen darting about the undergrowth.

SULPHUR-CRESTED COCKATOO
Cacatua galerita

GEOGRAPHIC RANGE: north, east and south-east.

HABITAT: varied vegetation types.

SIZE: 48–55 cm.

DIET: bulbs, fruit, grains, insects, insect larvae, nuts and seeds.

BREEDING SEASON: May to January, variable.

NEST: hollow in old-growth tree, lined with chewed wood.

EGGS: 1–6 white.

INCUBATION: 28–30 days.

FLEDGE TIME: 8–9 weeks.

VOICE: raucous screech.

GENERAL: The sulphur-crested cockatoo is a common and familiar bird throughout Australia. However, they can be a pest around urban areas and to crop farmers; they tend to form huge flocks that can descend and destroy entire crops.

© Dave Watts / ANTPhoto.com

TURQUOISE PARROT
Neophema pulchella

GEOGRAPHIC RANGE: south-east Qld; east NSW.

HABITAT: open forest.

SIZE: 20–22 cm.

DIET: seeds.

BREEDING SEASON: August to December.

NEST: tree and limb hollows, lined with chewed wood.

EGGS: 3–8 glossy, smooth, white.

INCUBATION: 20 days, mostly by the female.

FLEDGE TIME: 4–5 weeks.

VOICE: 'zwee-zwee-zwee'.

GENERAL: The turquoise parrot is distinguished by its bright electric-blue face and shoulders, which it flashes during courtship displays.

WOMPOO FRUIT DOVE
Ptilinopus magnificus

GEOGRAPHIC RANGE: east coast.

HABITAT: rainforest.

SIZE: 35–45 cm.

DIET: fruit.

BREEDING SEASON: June to January.

NEST: stick nest on ground, rock ledge, tree fork or vines.

EGGS: 1–2 glossy white to cream.

INCUBATION: 14–21 days.

FLEDGE TIME: 3–5 weeks.

VOICE: 'wallock-a-woo', or 'wom-poo', hence its name.

GENERAL: The wompoo fruit dove is perhaps the most beautiful of the Australian doves, with a white head, plum breast, green wings, and golden abdomen.

YELLOW-BREASTED BOATBILL
Machaerirhynchus flaviventer

GEOGRAPHIC RANGE: northern tip of Cape York, Qld.

HABITAT: rainforest.

SIZE: 11–12 cm.

DIET: insects.

BREEDING SEASON: August to April.

NEST: shallow suspended cup.

EGGS: 2 white, speckled.

INCUBATION: 14–18 days.

FLEDGE TIME: not known.

VOICE: high warbles and insect-like buzzing.

GENERAL: The yellow-breasted boatbill is a flycatcher with a wide bill, often seen foraging in the rainforest's middle storeys, where its prominent lemon-yellow breast catches the eye.

YELLOW-BILLED KINGFISHER
Syma torotoro

GEOGRAPHIC RANGE: northern tip of Cape York, Qld.

HABITAT: rainforest edges.

SIZE: 19–21 cm.

DIET: crabs, fish, frogs, insects, lizards, mice, snails, snakes and yabbies.

BREEDING SEASON: September to February.

NEST: tree hollows.

EGGS: 2–6 white.

INCUBATION: 21 days.

FLEDGE TIME: 3–5 weeks.

VOICE: loud descending trill, brief chirp.

GENERAL: The yellow-billed kingfisher is a stunning bird with its orange and green plumage and yellow bill.

YELLOW-BILLED SPOONBILL
Platalea flavipes

GEOGRAPHIC RANGE: most of the country except the central west.

HABITAT: freshwater wetlands.

SIZE: 76–92 cm.

DIET: fish, insects and molluscs.

BREEDING SEASON: variable.

NEST: sticks and vegetation in beds, bushes, islands or trees.

EGGS: 2–4 white.

INCUBATION: 26–31 days.

FLEDGE TIME: 7 weeks.

VOICE: deep reedy grunt, bill clattering.

GENERAL: A solitary bird, the yellow-billed spoonbill is often seen wading in fresh water.

YELLOW HONEYEATER
Lichenostumus flavus

GEOGRAPHIC RANGE: northern Qld.

HABITAT: forest, mangroves and gardens.

SIZE: 16–18 cm.

DIET: fruit, insects and nectar.

BREEDING SEASON: August to April.

NEST: cup of bark, cobweb, moss and sticks.

EGGS: 2–3 white to pink with blotches.

INCUBATION: 12–17 days by the female.

FLEDGE TIME: 11–20 days.

VOICE: varied whistles.

GENERAL: A melodious bird, the rich colouring of the yellow honeyeater can often be seen as it feeds on nectar in urban areas, as well as in forests.

PENGUIN BOOKS

Published by the Penguin Group
Penguin Group (Australia)
250 Camberwell Road, Camberwell, Victoria 3124, Australia
(a division of Pearson Australia Group Pty Ltd)

First published by Penguin Group (Australia),
a division of Pearson Australia Group Pty Ltd, 2005

10 9 8 7 6 5 4 3 2 1

Text copyright © Penguin Group (Australia) 2005
Photographic copyright remains with individual photographers

The moral right of the author has been asserted

All rights reserved. Without limiting the rights under copyright reserved above, no part of this publication may be reproduced, stored in or introduced into a retrieval system, or transmitted, in any form or by any means (electronic, mechanical, photocopying, recording or otherwise), without the prior written permission of both the copyright owner and the above publisher of this book.

Cover design by Adam Laszczuk & Claire Tice © Penguin Group (Australia)
Text design by Claire Tice © Penguin Group (Australia)
Cover photograph by Alamy Images

Printed in China by Everbest Printing Co. Ltd

National Library of Australia Cataloguing-in-Publication data:

Tropical birds of Australia.

ISBN 0 143 00393 3

1. Tropic - birds.

598.43

www.penguin.com.au